Franklin Lacerda de A. F. Júnior
David D. S. da Silva

Thermal and dynamic analysis and design of a brake system

Franklin Lacerda de A. F. Júnior
David D. S. da Silva

Thermal and dynamic analysis and design of a brake system

Brake Sizing, Structural Simulation, Thermal Simulation

ScienciaScripts

Imprint
Any brand names and product names mentioned in this book are subject to trademark, brand or patent protection and are trademarks or registered trademarks of their respective holders. The use of brand names, product names, common names, trade names, product descriptions etc. even without a particular marking in this work is in no way to be construed to mean that such names may be regarded as unrestricted in respect of trademark and brand protection legislation and could thus be used by anyone.

Cover image: www.ingimage.com

This book is a translation from the original published under ISBN 978-613-9-74575-3.

Publisher:
Sciencia Scripts
is a trademark of
Dodo Books Indian Ocean Ltd. and OmniScriptum S.R.L publishing group

120 High Road, East Finchley, London, N2 9ED, United Kingdom
Str. Armeneasca 28/1, office 1, Chisinau MD-2012, Republic of Moldova, Europe
Printed at: see last page
ISBN: 978-620-6-31322-9

Abstract: The different systems of a vehicle enable it to move safely, among these systems one of great importance is the brake system. The brake system promotes the stop of the vehicle efficiently and safely, for this to occur it is necessary to dimension it correctly, in order to avoid the occurrence of failures during its activation. During braking, the kinetic and potential energies of a vehicle are converted into thermal energy through the friction of the disc with the pad. The overheating of the discs can cause problems and lead them to fracture. Thus, the main objective of this study is to analyse the structural and heat dissipation capacity of the discs for a 250cc motorbike, ensuring that the brake does not fail during use. It was used
the ANSYS software for thermal and dynamic analysis in order to study whether the discs are able to perform all the braking without any damage occurring. The thermal analysis showed that the front disc will reach a maximum temperature of 259.56°C considered a good temperature, since the steel begins to have considerable variations in its resistance from 400°C, so the disc will not suffer the possible problems caused by overheating. The structural simulation of the rear disc showed a very high safety coefficient, allowing new studies and projects to be made regarding its geometry, and there may be a decrease in its mass by increasing the ventilation in the disc. The thermal analysis also shows a great result with a maximum temperature of 106.66 ° C, showing that the rear disc will not suffer from overheating problems.

Keywords: Brake sizing, structural simulation, thermal simulation.

SUMMARY

CHAPTER 1

INTRODUCTION

Vehicles are made up of different systems that together enable the vehicle to move and fulfil all the commands of the driver efficiently and safely.

These systems include the braking system, whose purpose is to brake the vehicle efficiently and safely, and which fulfils strict safety requirements.

Among these systems, the braking system can be highlighted, which aims to brake the vehicle efficiently and safely, in addition to being a function that meets strict safety requirements. Thus, the problem of this research deals with the development of brake discs that guarantee efficiency and safety during the braking process.

This study seeks the design of a brake system, with the aim that it performs its function effectively and efficiently, bringing safety to the pilot, making the structural and thermal analyses of the rear and front discs.

In order to perform the thermal sizing it is extremely important to stick to the defects that an overheating may cause in the brake discs, starting from this premise from the bibliographic studies it was found that the discs may suffer thermal fracture due to poor dissipation of the heat generated during braking due to the friction of the pads with the disc, this overheating may cause surface cracking due to thermal load, this occurs due to thermal shock and / or thermal fatigue, since the structural design depends mainly on the torque to which the disc will be subjected, and this torque may exceed the yield limit of the material.

The study is extremely important in the area of mechanical engineering since it deals with the design of a brake system, a system of great importance for a vehicle, even if it is a motorbike. The premises of the brake system design can be used in cars or large vehicles, also putting into practice all the learning absorbed during the course and encouraging the student to delve deeper into research.

As a methodological procedure, bibliographic research on the theme and competition regulations was used, in addition to the use of ANSYS *software* for structural and thermal analysis in order to study whether the discs are able to perform all the braking without any damage.

1.1 Evolution of the Brake System

According to Brembo *apud* Pinto (2014), the brake system dates back to the creation of the wheel. If it had not been invented, none of our current means of transport would exist. Historians believe that its emergence happened between 5000 and 4000 years ago, however the first wheel was not used for transport purposes, but in the construction of rudimentary tools. One of the first applications was probably the Potter's wheel, a device made to facilitate the manufacture of vessels to collect and heat water.

A long time passed and vehicles - those pulled only by animals, such as carriages - were fitted with real brakes, to reduce speed on slopes. Brake lever systems, belt brakes and then brakes with a wooden shoe that rubbed against the wheel. This type of brake is still used today in agriculture.

Scholars of this time did not expect the development of efficient brakes until the formulation of the laws of friction, still used by brake engineers today. The coefficient of friction, denoted by the letter *p,* represents a value, between 0 and 1, by which the applied normal force is multiplied. In the case of disc brakes, the coefficient between disc and pad is approximately 0.4, while in tyre-ground contact the coefficient of friction varies between 0.8 and 1 (BREMBO *apud* PINTO, 2014).

According to Brembo *apud* Pinto (2014), the creation of the brakes was mainly due to the invention of the wheel that needed some device to make its movement stop. The first brakes also have friction as a braking principle, this friction being caused by the contacts of the disc with the pad and the tyre with the ground. Thus, the braking system aims to reduce the speed of the moving vehicle progressively until it stops.

The brake system has the primary function of decelerating the vehicle, transforming its kinetic energy into thermal energy, dissipating it to the environment.

For this system to be efficient, it must present as characteristics resistance to erosion, light weight, high useful life, noise reduction and wear rate, but with an acceptable relationship between cost and performance (LIMPERT; BLAU *apud* LUCIANO, 2005).

Brakes are essentially dissipative devices and therefore generate a large amount of heat during their operation. Thus, they must be designed to absorb and transfer this heat without causing damage to themselves or their surroundings. Often, it is the ability to transfer heat of a device that limits its capacity, rather than its ability to transmit mechanical torque (NORTON 2004).

According to Limpert (1999), the deceleration caused to the vehicle by the brake system, causes the kinetic energy to be transformed into thermal energy, in which the thermal energy is dissipated to the environment. Thus, for the system to be efficient it is necessary that it has a satisfactory cost and performance ratio, taking into account the characteristics of the brake material such as mass, resistance to harmful environments and durability.

According to Boiochi *apud* Santos (2014), the brake system must have characteristics that can offer greater braking force, which allows good braking efficiency even in heavy vehicles, insensitivity to the friction coefficient of the pad, making the vehicle able to brake in a stable and constant way. Other important features of the disc brake are: 1) Higher braking force - this feature is generally used by heavy-duty vehicle manufacturers to increase brake performance and reduce braking distance. 2) Insensitivity to pad friction coefficient - the effect of this feature of the disc brake is good for vehicle stability, keeping the braking force constant.

However, it is necessary that it still presents some characteristics that define its project, among which:

- Thermal performance;
- Mechanical performance;
- Durability;
- Packaging;

- Pasta;
- Ease of construction;
- Ease of maintenance;

The disc brake system is basically composed of three basic components: the disc, the caliper, and the brake pads (GONÇALVES, 2004), as shown in Figure 1.

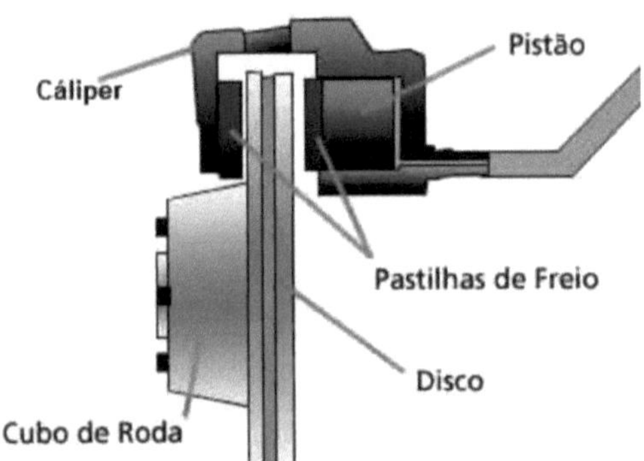

Figure 1- Components of a disc brake.

1.2 Brake caliper

The caliper is a vehicle device, fixed to its frame, and consists of one or more pistons that compress the pads on the brake disc. As for the drive, the caliper can be classified as pneumatic, employed in railway vehicles and some truck and bus brakes; hydraulic, used in most small commercial vehicles; or magnetic electric, by centrifugal effort. (JUNIOR, 2012). Regarding the way the pistons are housed in the caliper, it can be classified as floating Figure 2 (a) or fixed (b). Fixed calipers contain one or more pairs of pistons that act simultaneously on each side of the disc. In this way, they allow the pistons to receive the same pressure from the system (GONÇALVES, 2004).

To perform the locking of the brake is coupled a caliper, which has a cylinder that compresses the brake pads against the disc performing its locking. This system is usually used on the front axles of small vehicles, due to its higher manufacturing cost compared to the drum brake system, which is usually used on the rear axles (LIMPERT, 1999).

The caliper is fixed to the motorbike frame and has the purpose of supporting the pads, in the calipers are located the pistons that will transmit the pressure generated in the handle or pedal at the moment they are activated, causing the pad to generate friction with the disc.

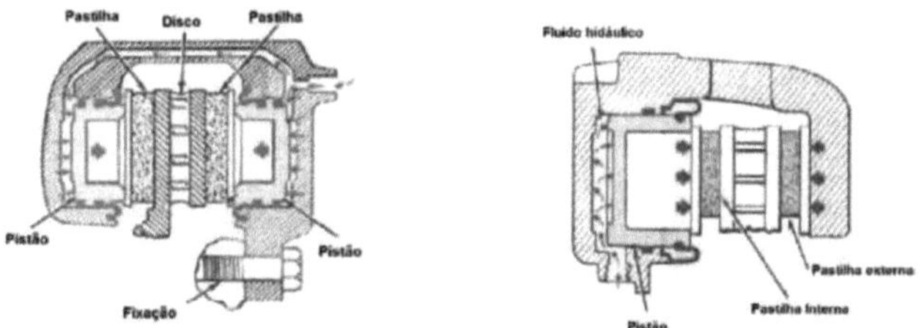

Figure 2- (a) Fixed clamp. (b) Floating clamp respectively.

Fixed calipers were replaced from the 1990s onwards by floating ones, mainly due to factors such as weight reduction, cost and physical space. The piston assembly is located on only one side of the caliper, applying the pressure of the system to the entire pad area. As soon as the friction material touches the disc, by reaction, it causes the outer pad to also contact the disc performing its braking (REHKOPF, HALDERMAN, PITTEL apud BORNHOLD, 2012).

1.3 Brake Pads

To facilitate maintenance, the friction material should be located on the pads, not on the discs. The pads, as shown in figure 3, are chemical compounds made up of several elements, these elements are made up of fibres and friction material in powder form. The specialised companies carry out several tests in the laboratory, to constitute the tablet of greater resistance, durability, and lower cost, present a coefficient of friction between 0.35 to 0.45 (LIMPERT, 1999).

According to Kruze (2009) brake pads are made of friction material and must be able to withstand and realise the transformation of kinetic energy into heat. Brake pads are produced using composite materials.

According to Pinto (2014), brake pads must withstand high temperatures, since kinetic energy will be transformed into thermal energy during braking, raising the temperature of the disc and pads, in addition to maintaining their mechanical properties unchanged even when subjected to such high temperatures.

Figure 3- Brake pad.

1.4 Brake disc

Brake discs suffer from a high torque, usually up to two or three times greater than that of the engine, causing an increase in the temperature of the discs, requiring a large cooling area. However, part of its surface is used as a friction surface at all times (JUNIOR, 2012).

The success of the disc brake is due to its low sensitivity to *fading* (when there is little heat dissipation in the disc, the pads reduce their coefficient of friction and the rider must apply the brake system more often). This system has excellent cooling, since its construction directly exposes the friction region of the disc to the air. Due to its manufacturing method, it is possible to design with a high thickness and thus add internal radial channels for ventilation. This type of construction, called a ventilated disc, allows the disc to work like a centrifugal fan, significantly increasing cooling efficiency. In discs, the effect of thermal deformations, which also cause *fading* in drum brakes, has no influence on the shape of the pad/disc contact region, since the disc is flat (ROSA; LEAL; NICOLAZZI, 2008).

According to the authors above, the brake discs are subjected to very high torques constantly which causes a sudden increase in temperature, so the brake discs for having their structure fully exposed to the environment allows this heat generated to be dissipated more easily. In addition to the possibility of using channels (drilling) in the discs further increasing the cooling efficiency.

The brake disc is usually made of cast iron, as it has a low coefficient of thermal expansion and a high coefficient of friction. It basically has two constructive forms: solid and ventilated. The solid disc is normally used in lower power ranges.

For the ventilated disc, the optimisation of the ventilation channels gives each disc a different shape Variations in the shape of the ventilation channels allow the optimisation of the air flow and therefore the cooling of the disc and these channels remove impurities from the pad material. This increases braking performance and prevents friction reduction due to heating (GONÇALVES, 2004).

According to Gonçalves (2004), the use of ventilated discs becomes more efficient due to the presence of ventilation channels through which the air passes, improving the cooling process, in addition to improving the cleaning process of the discs. Solid discs that do not have channels do not promote such efficient cooling and make it difficult to clean the discs.

Limpert (1999) emphasises that when the brake system reaches high temperatures (above 300°C), a loss of braking efficiency starts to occur due to the reduction of frictional forces between the friction material and the rotor, because of the reduction of the friction coefficient. This problem is called the *fade* effect.

The disc is the rotating element of the disc brake system. Being manufactured using cast iron, steel alloys, ceramics, etc. According to Abreu (2013) the disc surface, in addition to supporting the friction of the pad to reduce the rotation of the disc, ends up absorbing much of the heat generated during braking. Because they have an open structure, the air flow facilitates their cooling.

Figure 4 illustrates some alternative brake disc geometries. Figure 4 (a) represents a solid brake disc, which is the simplest and most widely used solution in passenger cars. Figure 4 (b) shows a ventilated brake disc, which allows a faster thermal exchange. Figure 4 (c) illustrates a ventilated brake disc with a grooved surface, which, according to Abreu (2013), allows better cleaning of the friction surface, elimination of water film and elimination of gases that arise due to the high temperatures generated on this surface.

The 4 (d) represents a ventilated brake disc with a perforated surface, which allows better gas exchange and waste output, due to the gases released and the particles generated by the brake pads when requested to the extreme, better water drainage when the vehicle travels on a wet track.

The brake discs shown in (b), (c) and (d) are used to optimise the cooling capacity and performance of brake discs and are therefore widely used to reduce speed as they have reliability, stability and the ability to promote high braking torques. Throughout braking, discs and pads lose their characteristics due to the increase in temperature that promotes the reduction of friction of the pads as well as overheating of the disc and brake fluid, in addition to causing effects such as the *fade effect*.

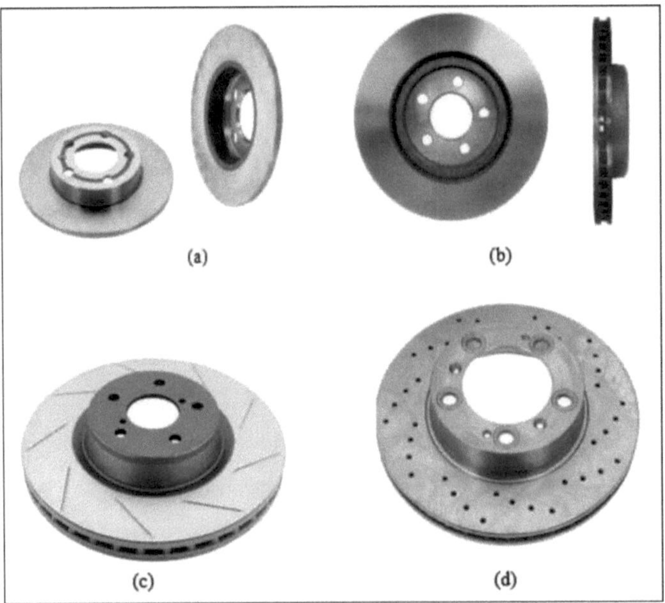

Figure 4- (a) Solid brake disc, (b) ventilated brake disc, (c) ventilated brake disc with grooved surface and (d) ventilated brake disc with drilled surface.

1.5 Finite Element Method

Physical models (usually on a reduced scale) and/or mathematical models are used to study the behaviour of physical systems. The advancement of science and the comparison between these models has led to a great development of mathematical models, providing realistic, reliable modelling and practical applications in engineering, much more economical than physical models (SORIANO, 2003).

Among the mathematical models, the finite element method was developed for the analysis of continuous media and can now be used to analyse all physical systems dealt with in engineering.

The finite element method (FEM) is one of the numerical methods for solving differential equations that describe many engineering problems. FEM, which originated in the field of structural mechanics, was extended to other areas of solid mechanics and later to other fields such as heat transfer, fluid dynamics and electromagnetism.

In fact, FEM has been recognised as a powerful tool for solving differential equations and integrodifferential equations and, in the near future, it may become the preferred numerical method of many areas of engineering and applied sciences. One of the reasons for its popularity is that the method results in naturally versatile computational programmes (KIM et al., 2011, P. 49).

For the analysis of the structure, the finite element method model is used, which represents the internal structure of all material, being very close to reality. According to Cesa (2010), the finite element method considers the region (continuum) of solution of the problem formed by small elements interconnected with each other. The region under study is modelled or approximated by a set of predefined discrete elements. Since these elements can be placed together in an uncountable number of different configurations, one can model quite complex geometric shapes.

The set of nodes and elements used for the discretisation of the model is called mesh. According to Budynas et al. (2006), the nodes are the fundamental entities of the elements, it is at the nodes that the elements connect to each other, where the elastic properties of the element are established and where the forces and boundary conditions are defined.

Through the use of polynomial functions integrated with matrices, the elastic behaviour of each element can be defined. According to Alves (2000), the main point for the elaboration of the CAE model is the choice of the elements that will represent each section of the structure.

According to Alves (2012), the study region is formed automatically, and can then be modified and modelled so that the properties of the elements can be established as detailed as possible, and each section of the structure can also be modified according to the need, so in more critical regions the number of nodes can be increased, thus improving the analysis.

The mesh is refined, i.e. the number of elements is increased, between analyses to increase the quality of the results in areas with large stress concentrations and zones of geometric transitions. Usually the results of the finite element analysis converge to the real result when the mesh is continuously refined, and consequently, the degree of the polynomial equations (P) used in the solution are increased (BUDYNAS et al., 2006). mesh is generated automatically by finite element *software* and usually during the first analysis a coarse mesh is used where no refinement area is defined. From the results it is then possible to identify the areas with the highest concentration of stresses and apply refinements in these regions until the results begin to converge to the exact result.

According to Alves (2012), in a first analysis the mesh is generated automatically being this mesh of low quality, from the analyses the regions of stress concentrations are established, and then a refinement can be made in the mesh in these regions making the results converge to reality.

1.6 Modelling method using ANSYS

ANSYS is a general purpose finite element modelling package for numerically solving a wide variety of mechanical problems. These problems include: static structural analysis / fluid dynamics acoustic and electromagnetic problems (both linear and non-linear), and heat transfer as well.

The acronym ANSYS stands for *Swanson Analysis Systems*, which translates to swanson analysis system, swanson in honour of the creator of ANSYS Jonh A. Swanson. There are two methods of using ANSYS. The first is through the graphical user interface or GUI. This method follows the conventions of popular Windows and X-Windows based programmes.

The second is via command files. The command file approach has a steeper learning curve for many, but has the advantage that an entire analysis can be described in a small text file, typically in less than 50 lines of commands. This approach allows for easy model modifications and minimal file space requirements.

The ANSYS packages used in the project were structural analysis and thermal transient, which allow us to solve complex problems on these topics, as well as easily connect to other physical analysis tools, providing even greater realism in predicting the behaviour and performance of more complex structures.ANSYS' suite of structural analysis software allows you to solve complex structural engineering problems and make better, faster design decisions. With finite element analysis (FEA) tools, you can customise and automate your simulations and parameterise them to analyse various design scenarios. ANSYS Structural Mechanics *software* easily connects to other physical analysis tools, providing even greater realism in predicting the behaviour and performance of complex products.

1.7 Brake System Structural Analysis

Basic brake system design is based on brake torque performance. Traction torque is measured in NM, i.e. a function of brake diameter. It is convenient to express the brake torque effectiveness of a single brake by a dimensionless measure, called the brake factor. The specific brake torque or brake factor, defined as the ratio of application to drag force, is a general indicator of the ability of a brake to produce torque on discs with different coefficients of friction (LIMPERT, 1999).

According to the author, the torque indicates the braking power capacity of the brake, which can be increased by increasing the friction coefficient of the brake. Caliper pads should have, in particular, disc brake pads should have porosity to minimise the effect of water on the coefficient of friction. These porous openings should not store contaminants such as salt or wear particles that affect friction.

The metallic components of the lining materials in contact with water will corrode the swept surface of the rotor. Although a vehicle that is used every day produces braking temperatures that evaporate any water, vehicles stopped in a moisture-rich environment may cause corrosion problems on the pads or disc. Electrochemical reactions penetrate the disc deep enough to change the basic surface of the disc. The results are brake noise and vibration that can only be corrected by installing new discs (LIMPERT, 1999).

According to Limpert (1999), the torque depends on the coefficient of friction of the disc with the pad and if contamination of the disc with water occurs, the torque will be compromised, so the geometry of the disc must be able to make all water flow easily through its surface.

Vibrations at the front end during braking operation are caused by fluctuations in braking torque, resulting from discs with non-uniform thickness or drums out of line. Variations in disc thickness are caused by a number of factors or maintenance defects, deposits of coating material on the rotor can cause metallurgical changes if subjected to extreme thermal conditions. Torque fluctuation can also cause loss of vehicle control (LIMPERT, 1999).

According to Limpert (1999), if the discs are subjected to high temperatures, changes may occur in their structure, altering the thickness of the disc, this change causes the fluctuation of the braking torque and may cause loss of control of the vehicle.

The braking torque can be understood as the "braking power" of the vehicle. More specifically, it is the measure of the friction force on the tyre (braking force on each axle, introduced generically as F) multiplied by the rolling radius of the tyre and the product of the sums of the moments of inertia and angular velocity of the rotating parts (SANTOS, 2014).

As we know when activating the brake system, the braking torque is produced, which will be responsible for stopping the vehicle, the torque can cause changes in the disc structure or even the rupture of the disc, being extremely important to predict if the designed disc will withstand the braking torque without suffering runout or rupture. In the structural analysis, in addition to determining the braking torque, it is necessary to perform tests or simulations to determine if the disc is suitable for use. Knowing that shear and normal stresses are produced during the braking torque, the failure criteria of the maximum shear theory and the von-mises failure criterion were used in the structural analysis.

The most common case of yielding of a ductile material, such as steel, is the sliding that occurs along the contact planes of the randomly ordered crystals that form the material itself. This sliding is due to shear stress. Shear stress acts on planes at 45° from the principal stress planes, these planes coincide with the directions of Luder's lines, indicating that rupture occurs by shear using the idea that ductile materials fail by shear, Henri Tresca proposed in 1868 his theory which is used to predict the failure stress of a ductile material subjected to any type of loading.

The yielding of the material begins when the absolute maximum shear stress reaches the value of the shear stress that causes yielding of the material when it is subjected to axial tension only.

Knowing that the Maximum Shear Stress Theory predicts that yielding begins whenever the maximum shear stress in any element becomes equal to or exceeds the maximum shear stress in a tensile test specimen of the same material when that specimen begins to yield. The maximum shear theory is also known as Tresca or Guest theory (BUDYNAS, NISBETT, 2011).

A material when deformed by an external loading tends to store energy internally throughout its volume. The energy per unit volume of the material is called strain energy density, this failure criterion is based on the distortions caused by strain energy. The strain energy density in a volume element of the material subjected to the three principal stresses, the failure criterion and issue is known as maximum distortion energy theory or von mises criterion.

Maximum distortion energy theory, is given by the following equations:

$$\tau_{max} = \frac{\sigma_1 - \sigma_3}{2}, \tau_{max} \leq \frac{\sigma_e}{2}.$$

The Von-Mises criterion is given by the following equation:

$$(\sigma_1 - \sigma_2)^2 + (\sigma_2 - \sigma_3)^2 + (\sigma_3 - \sigma_1)^2 = 2Y^2$$

Even though the maximum shear theory can provide a reasonable thesis for flow in ductile materials, it is necessary to use other methods to be sure of what is being investigated.

The maximum distortion energy has a better relationship with the experimental data and is therefore often even more requested. In order to be clear about the material's impairment the two tension tests will be performed.

1.8 Brake Disc Thermal Analysis

During braking, the kinetic and potential energies of a vehicle are converted into thermal energy through the friction of the disc with the pad. The energy of the vehicle during braking is transformed into thermal energy, so it is necessary to dissipate this heat so that there is no compromise of the brake disc (LIMPERT, 1999).

One of the functions of the brake is to store and/or dissipate the thermal energy generated at the interface. Since the structural integrity of a brake can be related to the temperature at the friction surface, most theoretical investigations are directed at determining the expected temperature increase during braking in a single braking or during repeated or continuous braking.

The results indicate that in the case of single braking the friction surface should be as large as possible to reduce the temperature. However for continuous braking, heat capacity and convective heat transfer are essential, i.e. the design parameters of the disc geometry differ for single braking and continuous braking. Theoretical investigations also indicate that for organic linings, of the heat generated during braking, approximately 95% is absorbed by the drum or disc and 5% by the brake lining or pad (KNOVEL S/A, 1976).

According to Knovel S/A (1976), the disc surface must be designed for the different types of braking, and must be able to absorb thermal energy, as this directly affects the disc structure, knowing that most of the heat will be absorbed by the disc and not by the pad.

Heat dissipation during braking is accomplished by cooling fluid around the brake surfaces. A study of the energy absorption capacity of various fluids shows that oil has an excellent capacity, and that air is reasonably close to oil. Water is also good, but its freezing characteristic makes it a poor candidate. An aesthetic mixture of water and ethylene glycol eliminates the freezing problem. But it has only one-third the energy absorption capacity of oil and half the energy absorption capacity of air as shown in Table 1 below (LIMPERT, 1999).

Energy Absorption Capacity of Various Fluids			
Fluid	**State**	**Energy Absorption Capacity**	
		Nm/kg	**ft.lb/lbm**
Air	Gas	221,126	74
Water	**liquid**	185,978	62,2
Oil	**liquid**	279,505	99,5
Mixture-Water and ethylene glycol	**liquid**	98,67	33

Table 1- Fluid absorption capacity

According to Limpert (1999), air and oil have about the same capacity to absorb the heat generated during braking, but water due to its freezing characteristic lags behind the former.

The heat transfer coefficient requires a minimum level of convection air over the disc surfaces. Consequently the blower must meet requirements for air flow as well as sufficient air pressure to keep the brakes free of contamination. The minimum air pressure required to push water off the disc depends on the depth of water the vehicle can travel through and the specific weight of the water or mud (LIMPERT, 1999).

According to Limpert (1999), for heat dissipation it is necessary that there is a minimum level of air on the surfaces of the disc, so that convection occurs, a minimum air pressure is also necessary to push the water that is on the disc, so that corrosion does not occur, so when designing the disc its geometry must be able to enable heat dissipation, cleaning of water and impurities.

The distribution of braking energy between the disc and pad cannot be readily predicted, the distribution of thermal energy between the pad and disc being directly related to the thermal resistance associated with both sides of the interface. Some difficulty may arise in determining the convective heat transfer coefficient of the disc. Studies on the heat transfer of a rotating disc have been carried out, however, the influence of the location of the caliper on the disc has not been incorporated in any theoretical analysis, radiation effects are neglected in most applications as they contribute only about 5 to 10% to the heat transfer in the disc (KNOVEL S/A, 1976).

According to Knovel S/A (1976), the distribution of thermal energy between the insert and the disc is difficult to predict, as is the determination of the convective heat transfer coefficient, so it is necessary in some cases to estimate this value with some approximation, and radiation can also be disregarded in the heat dissipation process.

Surface cracking due to thermal loading can occur as a result of two phenomena: thermal shock and/or thermal fatigue. Thermal shock exists when a single application of heat flux and subsequent cooling produces surface failure. Thermal fatigue exists when a series of cooling cycles results in surface failure. If there are temperature variations of sufficient magnitude on the disc surface, heat cracks will develop generally orientated in a radial direction.

One of the requirements for heat crack formation is that the thermal stresses exceed the elastic limit of the material, causing plastic deformations to form on the surface. In the subsequent cooling cycle, the original dimensions can no longer be obtained, thus producing tensile stresses that exceed the ultimate strength of the material (KNOVEL S/A, 1976).

According to Knovel S/A (1976), the increase in temperature in the disc can cause changes in its structure, which can cause thermal shock and / or thermal fatigue, these thermal effects result in plastic deformations or even failures on the surface of the disc, so it is extremely important to have an idea of the magnitude of the temperature to which the disc will be subjected, so that it does not suffer such damage in its structure.

CHAPTER 2

METHODOLOGY

ANSYS software was used as a methodological procedure for structural and thermal analyses in order to study whether the discs are able to perform all the braking without any damage occurring. Next, considerations will be made regarding the thermal and structural analysis in the brake system under study.

2.1 Structural Analysis - Braking Torque

The braking torque is given by Eq. (1)

$$T = F\,R_p + I_w \propto_w \qquad (1)$$

F being the braking force, given by the following eq. (2)

$$F = m\ x\ \mu_{pneu}\ x\ g \qquad (2)$$

The mass (m) is given by the sum of the mass of the motorbike and the mass of the rider. A minimum weight of *95 kg is* considered for the motorbike, taking it as a basis for stipulating the mass of the motorbike. For the weight of the rider we will use the average, by IBGE, of the population between 18 and 29 years, as shown in table 1 below:

18 years - 65.3 kg	***19 years - 65.9 kg***
20 to 24 years - 69.4 kg	***25 to 29 years - 72.7 kg***

Table 1- Mass of the Population According to IBGE

Thus, for a mass of *70kg, mt = 165kg*. However, the mass of the motorbike is not certain, so a safety coefficient of 1.1 is used. Therefore: m = 181.5kg.

Subsequently, the coefficient of friction of the tyre with the ground is calculated as $g_{P\,neu}$. This varies according to tyre condition, speed, soil type and ground conditions. For a new tyre and an asphalt without wear, it can be seen in table 2.

TYPE OF TRACK	P
Asphalt	0,6 a 0,95
Crushed stone	0,5 a 0,65
Dry land	0,5 a 0,7
Wetland	0,5 a 0,6
Sand	0,2 a 0,3
Snow	0,3 a 0,35

Table **2**- Tyre friction coefficient

Therefore, for the ideal conditions of the track we will consider g_{pneu} equal to 0.8.

The other parameter to be considered is g the acceleration of gravity which is equal to: g = 9.81 m/s^2

The next braking torque parameter to be calculated is the effective tyre radius, which varies from tyre to tyre.

Considering the use of Dunlop racing tyre *model* KR149 M *size*: 95/70R17 on the front wheel and Dunlop tyre *model* KR133 C *size*: 115/70R17 on the rear wheel, thus varying the effective radius of the front and rear tyres.

To calculate the effective radius, it is necessary to obtain the effective diameter of the tyre, which is obtained by adding the radius of the tyre and the section height, multiplied by a coefficient. This coefficient is 0.98 for radial tyres and 0.96 for diagonal tyres. The Dunlop tyre is built radially, so the coefficient of 0.98 is used.

For the calculation of the moment of inertia the tyre mass, the effective radius and the tyre radius, given by eq. (3), are required.

$$I = 1/2 \ \ x\,(\,m\,x\,(R^2 + r^2)) \tag{3}$$

The angular acceleration is given by the scalar acceleration and the effective radius. The acceleration (in this case deceleration) was obtained by considering a test track as a 350 m straight line, where a motorbike at 80 km/h or higher must be decelerated completely. Braking should start from 200 m of the track.

According to physics, we have eq.(4)

$$v^2 = v^2 + 2a\Delta S \tag{4}$$

Therefore, the angular acceleration is described by eq. (5)

$$\alpha_w = \frac{a}{R_p} \tag{5}$$

The braking torque can be calculated as described in the parameters.

2.2 Load Transfer

At the moment of braking there is a transfer of mass to the front axle called load transfer, which can be calculated from eq. (6) below:

$$s = \frac{W x \mu x H}{l} \tag{6}$$

Knowing that W is the total weight of the motorbike, μ the coefficient of friction, H the height of the centre of gravity to the ground and l the wheelbase. When the motorbike is static about 40% of its weight is on its front axle and about 60% on its rear axle, with the calculation of the load transfer one can then calculate the mass on the front and rear axle, which are given by the following eq. (7) and (8).

Mass on the rear axle:

$$mr = (W.\,0{,}6) - s \tag{7}$$

Mass on the front axle:

$$mf = (W.\,0{,}4) + s \tag{8}$$

2.3 Thermal Analysis - Braking Energy

During braking, the kinetic energy and potential energy of the vehicle are converted into thermal energy by the friction of the brakes. For a deceleration of the vehicle on a flat surface, from high speed *V1* to a low speed V2, the braking energy according to demonstration found in Knovel (1976) is given by:

$$E_b = \left(\frac{m}{2}\right) x\left({V_1}^2 - {V_2}^2\right) + \left(\frac{I}{2}\right) x\left({w_1}^2 - {w_2}^2\right) \tag{9}$$

If the vehicle undergoes a complete deceleration, we have that $V_2 = w_2 = 0$, then the previous equation will reduce to:

$$E_b = \frac{m\,{V_1}^2}{2} + \frac{I\,{w_1}^2}{2} \tag{10}$$

Knowing that m is the mass of the vehicle, V the velocity of the vehicle, W1 angular velocity and I the moment of inertia. When all the rotating parts are expressed in the revolution of the wheel we have that, $V = R.\,w$, therefore :

$$E_b = \frac{m(R.w^2)}{2} + \frac{I\,w^2}{2} \tag{11}$$

$$E_b = \frac{m}{2}\left(1 + \frac{I}{R^2 m}\right){V_1}^2 \tag{12}$$

$$E_b = \frac{K.m.{V_1}^2}{2} \tag{13}$$

K being the correction factor for rotating masses. Typical K values for passenger cars range from 1.05 to 1.15 at high speeds and from 1.3 to 1.5 at low speeds (LIMPERT, 1999).

2.4 Thermal Analysis - Braking Power

The braking power is given by eq. (14) below.

$$P_b = \frac{d(E_b)}{dt} \qquad (14)$$

When the deceleration of the vehicle is constant, its speed varies in time according to the following eq. (15).

$$V_{(t)} = V_1 - at \qquad (15)$$

Therefore the braking power will be:

$$Pb = k.\, m.\, a.\, (V1 - at) \qquad (16)$$

Inspections reveal that the braking power is not constant during the braking process. At the start of braking (t =0), the braking power is maximum, decreasing to zero when the vehicle is fully stopped.

The time t_s , for the vehicle to be stopped is given by eq. (17).

$$t_s = \frac{V_1}{a} \qquad (17)$$

Given that (a) is the deceleration of the vehicle.

The average braking power, can be calculated in thermal units, and is given by the following eq. (18).

$$q_0 = \frac{k(1-s).V_1.a.W.3600}{2.(778)} \qquad (18)$$

Knowing that s the slip of the tyre, it represents the energy absorbed by the tyre/road due to the partial slip of the tyre. For an extreme situation which would be the tyre locking, we have that s=1.

A vehicle that is being decelerated with its tyres operating close to their maximum braking capacity, without complete wheel lock-up occurring, will have the tyres operating at approximately 8 to 12% slip (ALEXANDRIA, 1976).

The acceleration a which in this case is deceleration has to be given as a function of g, so we have that ím/s^2 is equal to o.1g (LIMPERT, 1999). W is the weight supported by the axle, which will be different for the front and rear axles.

The maximum power produced during braking is equal to:

$$q_0 = 2.\ q_0 \quad (19)$$

CHAPTER 3

RESULTS AND DISCUSSION

3.1 Front Disc Structural Sizing

According to eq. (6) the charge transfer is calculated:

$$s = \frac{181{,}5x0{,}8x0{,}625}{1{,}25}$$

s=72,6kg

With the load transfer the front mass can be obtained from eq. (8): mf = (181.5xo.4) + 72.6 kg

mf = 145,2kg

With these values the front braking force is calculated eq. (2):

Ff = 145.2 x 0.8 x 9.81 = 1139.5N

As specified the front tyre is **Dunlop model KR149 M size: 95/70R17,** this nomenclature means that it is a tyre of inner diameter 17 inches and outer 509mm, 88mm wide and 70% of the width value is the height of the tyre section, thus 61.6mm. According to the consulted catalogue its weight is 2.73 kg. Therefore, the effective diameter of the tyre can be calculated as specified:

Dp = [509 + (61.6 x 2)] x 0.98 = 619.6mm.

Therefore the effective tyre rim is Rp = 309.8 mm.

With the effective radius of the tyre, the moment of inertia is calculated according to eq. (3):

$$I = \frac{1}{2} x\ (2{,}73\ x\ (0{,}309^2 + (0{,}254^2))$$

$$I = 0{,}218\ kg\ .\ m^2$$

After calculating the moment of inertia it is necessary to calculate the deceleration equation (4):

The acceleration (in this case deceleration) was obtained for a test track which is a 350 m straight line, where a motorbike at 80 km/h or higher must be completely decelerated. Braking should start from 200 m of the track.

$$a = 21{,}34\ m/\ s^2$$

With the effective radius and the deceleration, the angular acceleration is calculated using eq. (5):

$$\propto_w = \frac{21{,}34}{0.309} = 69{,}25\ m/rad$$

Finally the front braking torque is calculated eq. (1):

$$Tf = 1139.5x0.309 + 0.218x69.25 = 367.2\ N.\ m$$

3.2 Front Disc Thermal Sizing

The calculation of the braking power is given by eq. (18)

$$q_0 = \frac{1{,}05\ x\ (1 - 0{,}12)\ x\ 22{,}22\ x\ 2{,}1\ x\ 1452\ x\ 3600}{1556} = 144842\ \frac{btu}{h}\ \text{ou } 42438{,}7W$$

With the calculation of the braking power, the maximum braking power is calculated using eq. (19)

$$q0 = 84877.4\ W$$

3.3 Rear Disc Structural Dimensioning

According to eq. (6) the charge transfer is calculated:

$$S = \frac{181,5x0,8x0,625}{1,25}$$

s =72,6kg

With the load transfer the back mass can be obtained from eq. (7):

mr = 108,9 - 72,6 = 36,3 kg

With these values the rear brake force is calculated eq. (2):

Fr = 36,3 * 0,8 * 9,81 = 284,8824 N

As specified the rear tyre is **Dunlop *model* KR133 C *size*: 115/70R17**. This nomenclature means that it is a tyre with an inner radius of 17 inches and an outer radius of 512mm, 115 mm wide and 70% of the width value is the height of the tyre section, so 80.5 mm. According to the consulted catalogue its weight is 3.9 kg. Therefore, the effective diameter of the tyre can be calculated as specified:

$$Dp = [512 + (80.5 \; x \; 2)] \; x \; 0.98 = 659.5 \; mm.$$

Therefore the effective tyre rim is Rp = 329.77 mm.

With can of the effective radius of the tyre the moment of inertia is calculated eq.(3):

$$I = \frac{1}{2} x \; (3,9 \; x(0,329^2 + 0,256^2))$$

I = 0,338 kg . m2

The deceleration is the same as calculated internally for the front tyre, with the effective radius and the deceleration the angular acceleration is calculated eq. (5):

$$\propto_{w} = \frac{21{,}34}{0.329} = 64{,}86 \text{ m/rad}$$

Finally the front braking torque is calculated eq. (1):

$$Tr = 284.8824x0.329 + 0.338x64.86 = 115.65 \text{ N. m}$$

3.4 Rear Disc Thermal Sizing

The calculation of the braking power is given by eq. (18):

$$q_0 = \frac{1{,}05\ x\ (1 - 0{,}12)\ x\ 22{,}22\ x\ 2{,}1\ x\ 363\ x\ 3600}{1556} = 36210{,}5\ \frac{btu}{h}\ \ ou\ 10609{,}7\ W$$

With the calculation of the braking power, the maximum braking power is calculated using eq. (19)

$$q_0 = 21219.4\ W$$

3.5 Structural Simulation Front Disc

After the disc and its geometry were sized, a static structural analysis of its flow was made in the ANSYS software, placing the fixed points, that is, the points where the disc will be fixed to the wheel, and the torque to which it will be subjected, the Von-Mises criteria and maximum shear were used, as well as the safety coefficient, below are the simulation results.

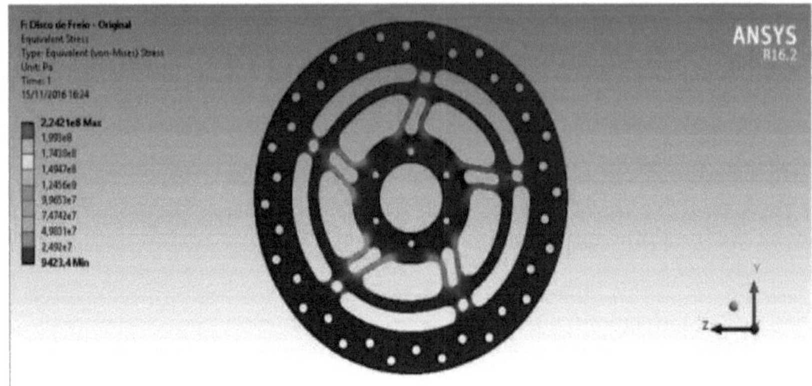

Figure 6- Front disc analysed by the Von-Mises criterion

According to the Von-Mises criterion the maximum stress the disc will reach is 224.2 MPa, below the yield strength of the material which is 250 MPa.

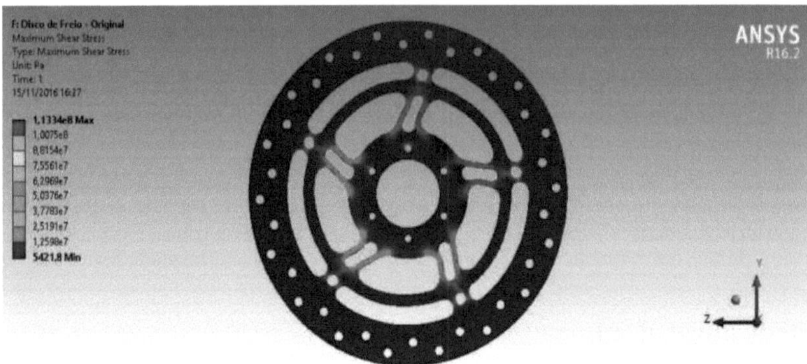

Figure 7 - Front disc analysed by maximum shear criterion

According to the maximum shear theory the maximum stress that the disc will be subjected to is 113.34 MPa, below the yield strength of the material which is 150 MPa.

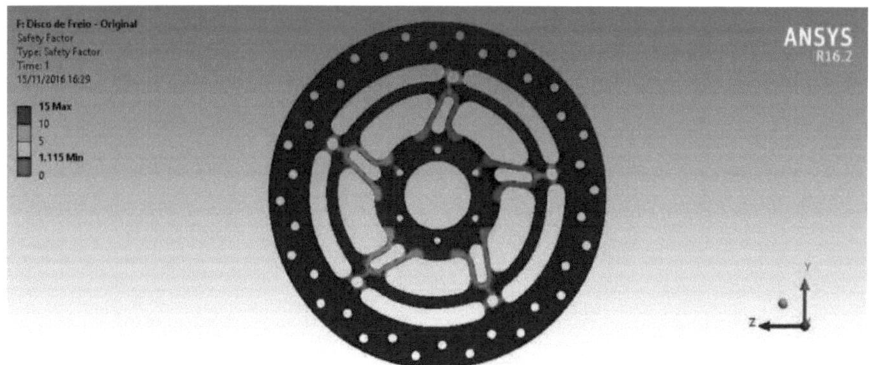

Figure 8 - Front disc safety factor

Analysing the safety coefficient, we found that it has a safety coefficient of 1.11. In order to increase the safety coefficient of the disc, a change was made in the region of greater stress concentration in it, this change increased some of the mass of the disc since it was necessary to fill some spaces in the geometry. Below are the data from the analysis of the modified disc.

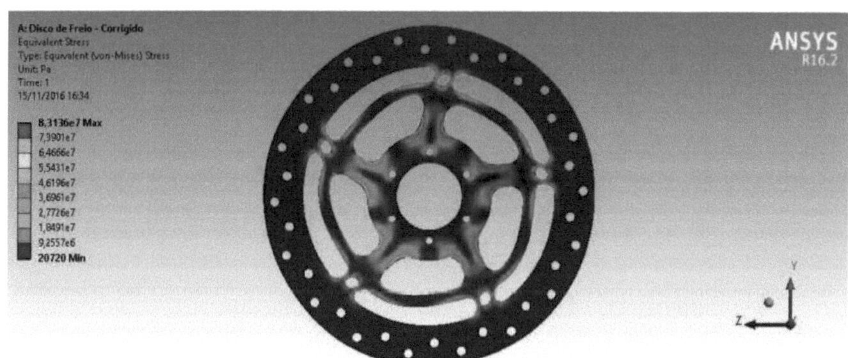

Figure 9 - Corrected front disc analysed by the Von-Mises criterion.

After the modification the maximum stress according to the Von-Mises criterion was 83.1 MPa.

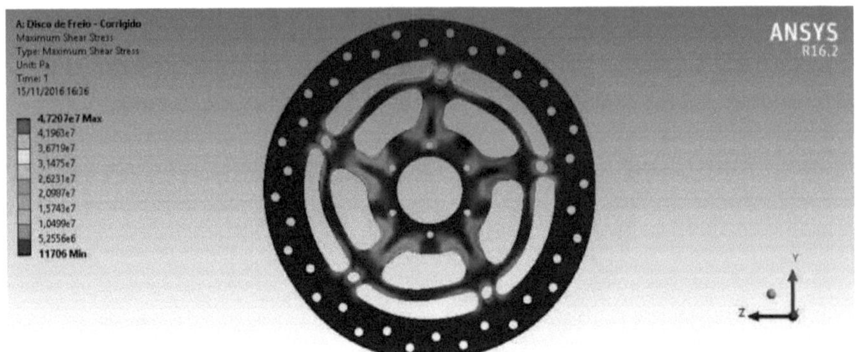

Figura 10 - Corrected front disc analysed for maximum shear

After the modification the maximum stress according to the maximum shear criterion was 47.20 MPa.

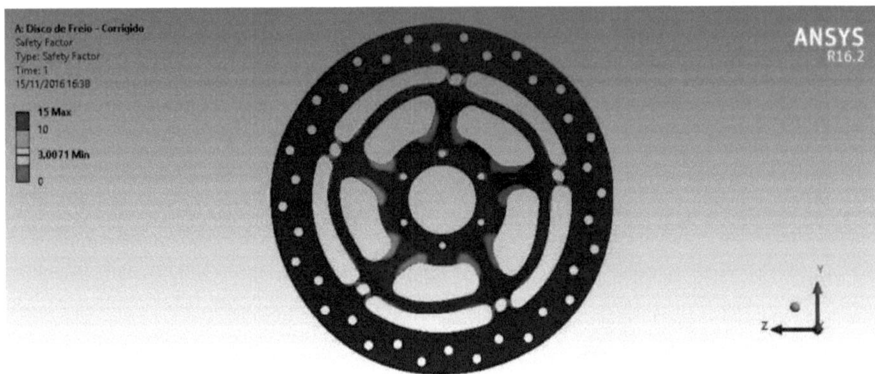

Figura 11 - Corrected front disc safety factor

With the modification the safety coefficient rose to 3, i.e. there was a significant improvement in the safety of the disc.

3.6 Front Disc Thermal Simulation

In the thermal analysis of the disc, the dissipation of this heat by convection and radiation was considered, the main variable of convection being the heat transfer coefficient by convection, which depends on the air speed, its direction, the temperature of the body and the geometry of the disc, and can be different for different points of the same surface, so the definition of this coefficient becomes difficult since the characteristics to which it depends change constantly.

For the simulation, three values of coefficients that vary in the time in which braking occurs were used, the first for forced convection and the last two for natural convection, respectively 35, 25 and 17 [w/m^2 .k], emissivity is the most important parameter of radiation for simulation and the emissivity of steel was considered 0.66.

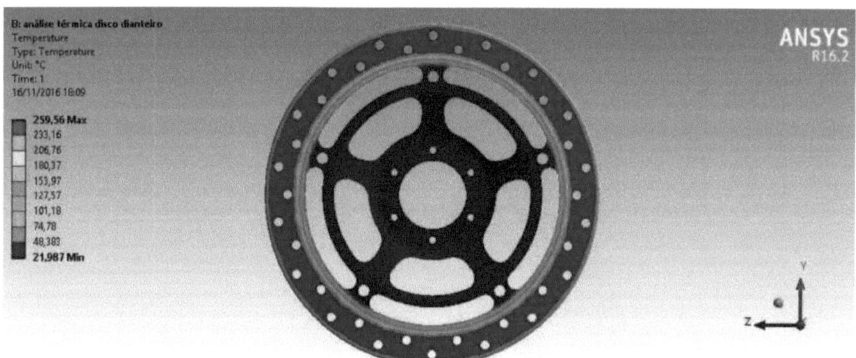

Figure 12 - Front disc thermal analysis

From the thermal analysis it was found that the maximum temperature that the disc will reach is 259.56°c, in the region where the insert is in contact with the disc.

3.7 Rear Disc Structural Simulation

Figure 13 - Rear disc analysed by the Von-Mises criterion

The maximum stress in the rear disc according to Von-Mises will be 18.14Mpa, below the yield strength of the material.

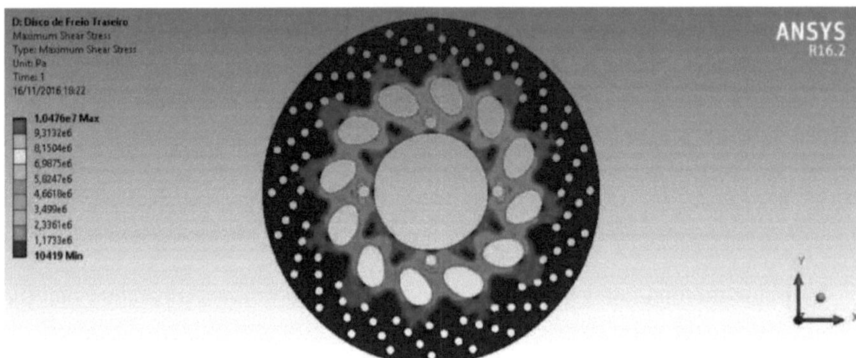

Figure 14 - Rear brake disc analysed by maximum shear

The maximum stress in the rear disc at maximum shear will be 10.47Mpa, below the yield strength of the material.

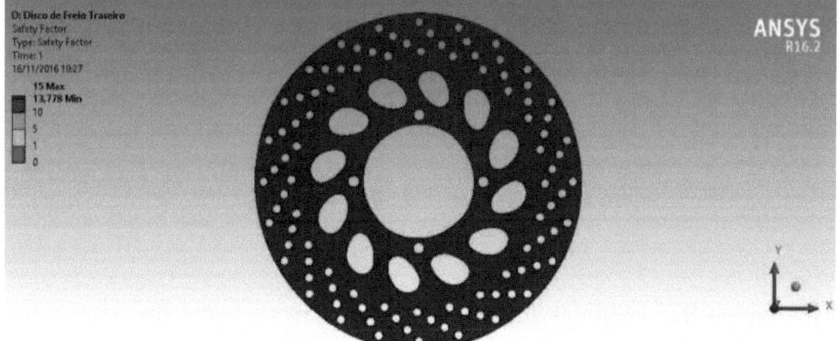

Figure 15 - Rear disc safety factor

The safety factor of the rear disc was found to be 13.77.

3.8 Rear Disc Thermal Analysis

The thermal simulation of the rear disc was performed using the same criteria as the front disc simulation.

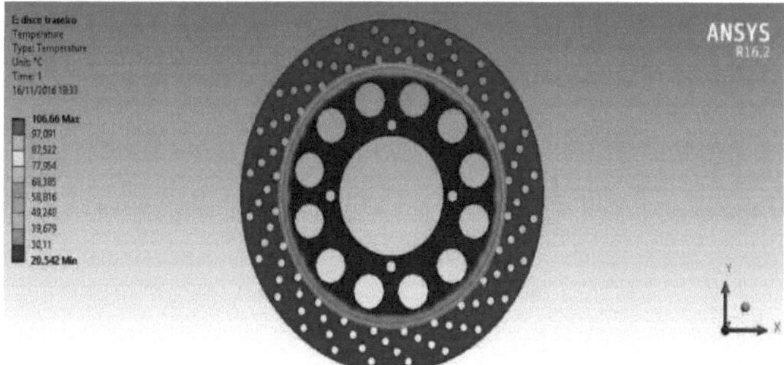

Figure 16 - Rear disc thermal simulation

From the thermal analysis it was found that the maximum temperature that the disc will reach is 106.66°c, in the region where the insert is in contact with the disc.

CHAPTER 4

CONCLUSION

The analyses made in the ANSYS *software* served as an aid in the feasibility of the construction of the discs, as it allows the prediction of their failures, also reducing costs, since the simulation is much cheaper than the development and construction of prototypes.

J Regarding the structural simulation of the front disc, the first disc to be developed shows a safety coefficient of only 1.1, showing the need to change the structure of the disc to increase this coefficient. For this correction, the spaces in the stress concentration region were filled and, with the new geometry, the coefficient rose to 3, an optimal value of safety coefficient since a disc with an ideal safety coefficient of 2 can be considered;

J The thermal analysis showed that the disc will reach a maximum temperature of 259.56°C, a good temperature, since the steel begins to have considerable variations in its resistance from 400°C, so the disc will not suffer the possible problems caused by overheating.

J The structural simulation of the rear disc showed a very high safety coefficient of 13.77, allowing new studies and projects to be carried out regarding its geometry, and there may be a decrease in its mass by increasing the ventilation in the disc. The thermal analysis showed a maximum temperature of 106.66°C, showing that the rear disc will not suffer from overheating problems.

Finally, it is concluded that the results indicate that the designed discs are suitable for use without any structural or thermal failures occurring in the discs.

CHAPTER 5

REFERENCES

ALVES, Renan Tassinari, Static structural analysis of the discharge system of a harvester using ansys software, FAHOR Faculdade Horizontina, Horizontina, 2012.

ABREU, R. M., 2012, "Simulation and Testing of Automotive Brake Mechanism". Master's Dissertation - Postgraduate Programme in Production Engineering, Federal University of Minas Gerais, 137p.

BREMBO, S.P.A.; 1997, "The Brake Disc Manual". Manual, Curno, 128p.

BUDYNAS, G. RICHARD; NISBETT, KEITH J.; Shigley's Mechanical Engineering Design. 8ed. United States: McGraw-Hill Primis, 2006.

BOIOCCHI, T. "Technological differences between tractors, trailers and impact in the safaty and drivability". In: Coloquium Internacional de Freios, 4, Caxias do Sul, 1999, p. 23-28.

BORNHOLD, Adelson Moacir. Dimensioning of a brake system for small Off Road vehicle, Fahor-Horizontina College, 2012.

CESA, T. R. Design of a rollover protection structure for agricultural tractors using computer simulation. 2010. Dissertation (Master in Design) - Faculty of Architecture, Federal University of Rio Grande do Sul, Porto Alegre, 2010.

GONÇALVES, J. Optimisation of friction material parameters of brake systems using genetic algorithms. 2004. Thesis (PhD in Mechanical Engineering) - Federal University of Santa Catarina. Florianópolis 2004.

GRADELA, Fabíola Bailão, Identification of failures using FMEA in the brake system of on-road racing cars - FÓRMULA SAE, University of São Paulo-USP, 2013.

HALDERMAN, J. D. Automotive Technology. 4. ed. New Jersey: Prentice Hall, 2012.

INFANTINI, M. B., 2008, "Performance Variables of Brake Systems", Master's Thesis, School of Engineering - Postgraduate Programme in Mechanical Engineering - Federal University of Rio Grande do Sul, Porto Alegre - RS, Brazil.

JUNIOR, A. A. dos S. Brakes and clutches by friction (2012). Available at: <www.fem. unicamp.br/~lafer/em718/arquivos/FreiosEmbreagens.doc>. Accessed on Aug. 2016

KAWAGUCHI, H. (2005) Comparison of Subjective vs. Objective Braking Comfort Analysis of a Passenger Vehicle. Professional master's thesis. Polytechnic School, University of São Paulo. 101p.

KIM, NAM-HO; SANKAR, BHAVANI V.; Introduction to Finite Element Analysis and Design. Translated by A. E. A. Kurban. Rio de Janeiro: LTC, 2011.

KNOVEL S/A. Engineering Design Handbook. Analysis and Design of Automotive Brake Systems, USA, 1976.

KRUZE, G. A. S., 2009, "Evaluation of the Coefficient of Friction in a Reduced Scale Dynamometer", Master's Thesis, School of Engineering - Postgraduate Programme in Mechanical Engineering - Federal University of Rio Grande do Sul, Porto Alegre - RS, Brazil.

LIMPERT. R. Brake Design and Safety, 2ª ed., USA, 1999.

LOMBRILLER, Silvia Faria. Thermal and Dynamic Analysis of the Disc Brake System of Heavy Commercial Vehicles, São Carlos School of Engineering, University of São Paulo, 2002.

LUCIANO, M. A. Reuse of information and knowledge to support the design of friction material. 2005. Post-Graduation (Production Engineering) - Federal University of Santa Catarina. Florianópolis 2005.

NORTON, R. L. (2004) Machine Design. Publisher: Bookman. 2nd edition. P 817 - 834.

PINTO, Rafael Lucas Machado, Analysis of variables that influence vehicle braking performance through tests based on the KRAUSS methodology, Federal University of Minas Gerais, 2014.

PITTEL, F. M. M. The influence of residues and cooling time of the brake disc on the friction behaviour in braking. 2011. Monograph (Mechanical Engineer) - Federal University of Rio Grande do Sul. Porto Alegre 2011.

SANTOS, Gustavo Carvalho Martins dos. Design and dimensioning of a brake system applied to a SAE formula vehicle, Federal University of Rio de Janeiro, Polytechnic School, 2014.

SORIANO, HUMBERTO LIMA. Finite Element Method in Structural Analysis. São Paulo: Edusp, 2003.

SHARP, B., 2013. "Fading, The Enemy of the Driver." Available at http://autoentusiastas.blogspot.com.br/2011/06/fading-o-inimigo-do-motorista.html Accessed Oct. 2016.

SHIGLEY, J.E. Elementos de Máquinas, 8ª ed., AMGH, São Paulo, 2011.

ROSA, da E.; LEAL, L. da C. M.; NICOLAZZI, L. C. An introduction to quasi-static modelling of wheeled motor vehicles (2008). Available at: < http://www. grante.ufsc.br>. Accessed on: November 2016.

MOORE, JUANW . Available at :< http://www.uff.br/petmec/downloads/resmat/W%20-

%20Apendice%20C%20Materiais2.pdf>. Accessed on: November 2016.

MECHANICAL WORLD. Available at: <http://www.mundomecanico.com.br/wp-content/uploads/2012/02/T ermografía.pdf> Accessed Aug. 2016.

CORRÊA, FELIPE. Available at:

<https://www.google.com.br/search?q=propriedades+mec%C3%A2nicas+do+a%C3%A7o&e

spv=2&biw=1242&bih=602&source=lnms&tbm=isch&sa=X&ved=0ahUKEwjfipj0i

LHQAh
WDipAKHYZHBrYQ_AUIBigB#tbm=isch&q=coefficient+of+convec%C3%A7%C
3%A3o &imgrc=JO9NpPN-xL5fiM%3A, Prof. M. Sc. Felipe Corrêa 20 November 2015, PUC Goiás>. Accessed on Aug. 2016

ENGINEERING CIVILISATION. Available at:

<https.construcaocivilpet.wordpress.compage3> Accessed Oct.. 2016

ANSYS, INC. Available at: <http://www.ansy s.com/products/structures> Accessed Nov 2016

ANSYS, INC. Available at: <http ://www.ansys. com/products/fluids/heat-transfer> Accessed Aug. 2016.

Printed by Books on Demand GmbH, Norderstedt / Germany